Explorers
What Lives in the Oceans?
Ocean Garbage "Patches"

Three Informational Texts About Science at Sea

by Jeanette Leardi

Table of Contents

Focus on the Genre: Informational Texts 2

Tools for Readers and Writers. 4

The Ocean: The Last Frontier 5

Explorers of the Deep 6

What Lives in the Oceans? 16

Ocean Garbage "Patches" 22

The Writer's Craft: Informational Texts 30

Glossary 32

Make Connections Across Texts . . . Inside Back Cover

FOCUS ON THE GENRE

Informational Texts

What is an informational text?

Informational text is nonfiction text that presents information in an accurate and organized way. It is often about a single subject such as an event or time period in history or a scientific discovery. It may be about any topic, such as a sport or a hobby. The research report that you write for a school assignment is an informational text. So is an article you read in your favorite fashion magazine or on a Web site. A newspaper account of a local election and a history book chapter on a famous battle are additional examples of informational texts.

What is the purpose of informational texts?

Informational text has one main purpose: to inform. The best informational writing does this in a way that keeps readers' attention. It pulls readers in, making them want to keep reading and to know more about the topic.

How do you read an informational text?

When you read an informational text, look for facts and for the details that support them. Read critically to make sure conclusions make sense. If there are different ways to look at an event or situation, make sure they are given. Ask yourself: *Did I learn something new from this text? Do I want to know more about it? Can I draw my own conclusions from what I have read?*

Features of an Informational Text

- The text has a strong beginning that hooks the reader.
- The information is accurate, and the facts have been checked.
- The text uses primary sources when appropriate.
- The information includes graphics that support the text.
- The text has a logical organization of major concepts.
- The text includes multiple perspectives so that a reader can draw his or her own conclusions.
- The text has a strong ending that keeps readers thinking.

Who writes informational texts?

Writers who know their topic well write good informational texts. They do this by becoming mini-experts on the subjects they are writing about. They make sure that they support the information in their work with historical facts, scientific data, graphics such as time lines and diagrams, and expert evidence. They provide more than one person's point of view. They use primary sources, firsthand information like journals and photographs.

Tools for Readers and Writers

A Strong Ending

A strong ending is the author's last chance to engage readers. Strong endings for informational texts might include a summary of the text, a restatement of facts, or an observation. More importantly, it calls for readers to think, or rethink, their thoughts about the text's subject matter.

Descriptive Language: Word Origins

Where do English words come from? Did someone wake up one morning and decide to call things having to do with water **aquatic**? No. Most English words come from other languages such as Greek, Latin, German, and French, to name a few.

The word **aquatic** comes from the Latin word for water, *aqua*, and the suffix *-ic*, which means "relating to." Readers who know the meaning of a word from another language can usually transfer that information to unknown words and build vocabulary.

Comprehension: Make Inferences

Writers don't explain everything in a story. Often they provide clues and evidence in their texts and expect the reader to read between the lines, or make inferences. Good readers consider information the writer provides, then think about other truths the information suggests. To make an inference, look for parts of the text that make you stop and think to yourself, *I wonder if the writer is saying that . . .*

The Ocean: The Last Frontier

We know a great deal about Earth's seven continents. For centuries, people have explored the deserts, jungles, and icy lands at both poles. Photos and videos of Earth have been taken from the ground and from outer space. We know the dimensions of rivers and mountains. We even know the shapes of the continents down to the shortest shoreline.

But a huge part of Earth remains a mystery. And that part makes up more than 70% of the planet! It's Earth's oceans. According to ocean photographer and conservationist Eric Cheng, scientists have explored only 5% of the world's oceans. There is still a huge amount of water to explore and innumerable things to learn about it. "The oceans are Earth's last frontier," Cheng says.

How are scientists exploring this frontier? And what are they discovering about it? Read on and make some discoveries of your own as you learn about science at sea!

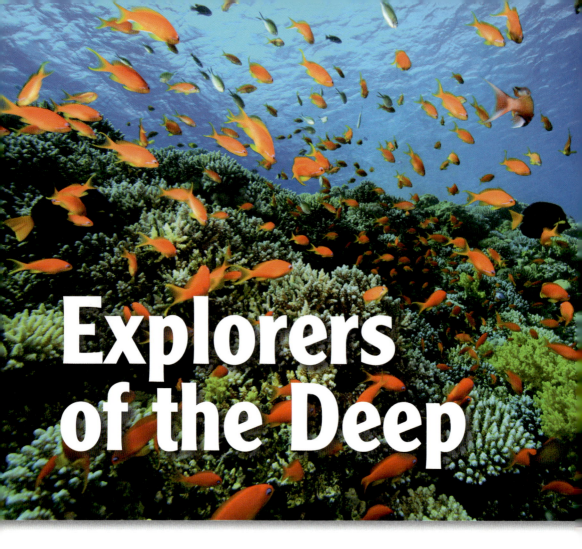

Explorers of the Deep

Ashanti Johnson wants to clean up the oceans. She is studying pollution off the coast of Georgia in the United States. Nancy Knowlton is on the other side of the world. She is counting the fish in a Pacific coral reef. Laurence Padman is more than 6,000 miles south of Knowlton. He is mapping the shape of the ocean floor beneath Antarctica. And just about anywhere in the world, Robert Ballard can be found exploring sunken boats and underwater volcanoes.

These four people are doing four very different kinds of work. But they are alike in one way: They all study the ocean. Their work is very important to the survival of Earth.

Studying Ocean Life

Nancy Knowlton is a **marine** biologist. She studies the living things in the seas and oceans. This includes plants and animals. Some of these living things can be microscopic, or too small to see with one's eyes. Others, such as whales, are huge.

Knowlton wants to know how these creatures live together. She frequently studies coral reefs. Coral reefs are stretches of corals, undersea **organisms** that look like rocks but are really animals. These reefs are home to one-quarter of all sea life! They also form barriers that protect islands from big waves and hurricanes. Knowlton is concerned that coral reefs are dying off because of people. Fishermen who catch too many fish cause an imbalance in ocean life. Companies that dump waste into the oceans poison sea creatures. Knowlton says, "I think you have to make people realize just how incredibly serious the situation is, but also that there's something they can do about it."

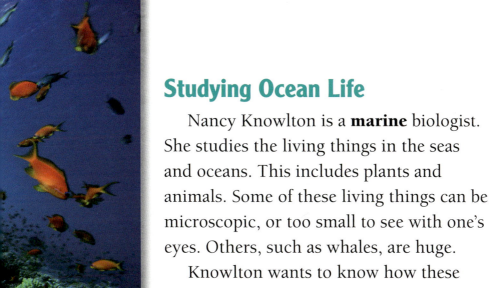

a reef in the Philippines with dead coral

Ashanti Johnson would agree with Knowlton. She, too, sees how pollution is hurting sea life. Johnson is a chemical oceanographer. She measures the poisons that find their way from rivers to the ocean. She knows that when small fish eat the poison, they can get sick and die. But those same small fish can be eaten by larger fish. Then the larger fish, too, get sick and die. People who eat either fish are in danger as well. Johnson looks for the sources of these poisons. They may come from factories or power plants. She shares her research with the government. Then the government works with the factories and the plants to clean up the waters.

Oceanographers study the pollution poisoning fish in lakes, rivers, and oceans.

Peering into Ocean Depths

Where Laurence Padman works, the water is clean and clear. But it is incredibly cold and deep. It is so cold and deep that scientists can't explore much of what is underwater.

EXPLORERS OF THE DEEP

Laurence Padman

elephant seal

Padman is exploring the waters around Antarctica. He is working with a team of oceanographers to map the Antarctic sea floor. Knowing the shape of the ocean floor will help scientists better understand Antarctica's water temperatures and movements, or currents. Scientists can then understand how Antarctica's sea life is affected by its climate.

If the water is so cold and deep, how does Padman do his research? He has unusual help: elephant seals. The seals can swim throughout the deep Antarctic seas. They can go where divers can't. Padman attaches a tag to the head of each seal. The tags record the water temperature, pressure, and depth. Padman has learned that the Antarctic waters are a lot deeper than scientists thought!

Explorers of the Deep

Robert Ballard knows all about exploring ocean depths. He is a geological oceanographer. He is interested in the rock formations under the sea. Ballard has made more than 120 research trips. He has traveled around the world and he has discovered amazing things. He found sea creatures in places where scientists thought nothing could live. He found underwater volcanoes. But his discoveries aren't just of natural things. In 1985, Ballard was the first oceanographer to discover the wreck of a famous ship called the *Titanic*. Ballard and his crew still hunt for shipwrecks. They have found boats, clay pots, stone statues, and metal coins. Some of those objects are more than 2,000 years old! Ballard says, "The deep sea is the largest museum on Earth. It contains more history than all other museums on Earth combined." He wants to keep exploring all the hidden history that the oceans contain.

Robert Ballard discovered the wreck of the *Titanic*.

Jacques Cousteau   *Calypso*

Pioneer of Oceanography

Knowlton, Johnson, Padman, and Ballard keep making important discoveries. All of them have one person to thank: Jacques Cousteau. He was a great **pioneer** of oceanography.

Jacques Cousteau was born in France in 1910. When he was twenty-six, he got the chance to swim underwater wearing goggles. A whole new world opened to him. In 1950, he bought an old war ship. He turned it into a floating science laboratory, the first of its kind. Cousteau named the ship *Calypso*. For forty years, Cousteau and his crew sailed the *Calypso*. They explored many places. They filmed their ocean-going adventures and made 115 television programs. Cousteau wrote dozens of books about the ocean world. In 1974, he founded the Cousteau Society to teach people about sea life. The Society also raises money for ocean research.

In 1985, the United States honored Cousteau. He was awarded the Presidential Medal of Freedom. Jacques Cousteau died in 1997, but his work lives on. He said, "The sea, once it casts its spell, holds one in its net of wonder forever." Cousteau was speaking about himself, but he might as well have been speaking for Knowlton, Johnson, Padman, and Ballard.

In fact, his words speak for all oceanographers everywhere.

Want to Be an Oceanographer?

An oceanographer is a scientist who studies the ocean. There are four kinds of oceanographers. One kind focuses on life science, or biology. Another deals with earth science, or geology. Still another studies the **physical** energy in the oceans. And yet another is concerned with the basic materials, or chemicals, of the sea. Each kind of oceanographer helps all scientists better understand the mysteries of Earth's seas and oceans.

Being an oceanographer is often quite adventurous and fun. But it also requires a lot of hard work. Special training is needed. And oceanographers must have certain personal qualities.

Oceanographers must know a lot of science and math and how to work with computers. They must pay attention to details and enjoy analyzing data. Oceanographers also have to be good communicators. They need to write reports about their discoveries for other scientists and students.

Ashanti Johnson is a chemical oceanographer.

Oceanographers need other skills, too. They must be curious about the world. They should care about the health of living creatures and the entire planet. Oceanographers rarely do their exploring and researching alone. So they need to be able to work well with others. A project can take years to complete, so oceanographers need to be patient. And, of course, they should love to swim and dive.

A scuba diver (right) explores the Indian Ocean. Researchers aboard a nuclear submarine (below) look for geothermal vents in the Atlantic Ocean.

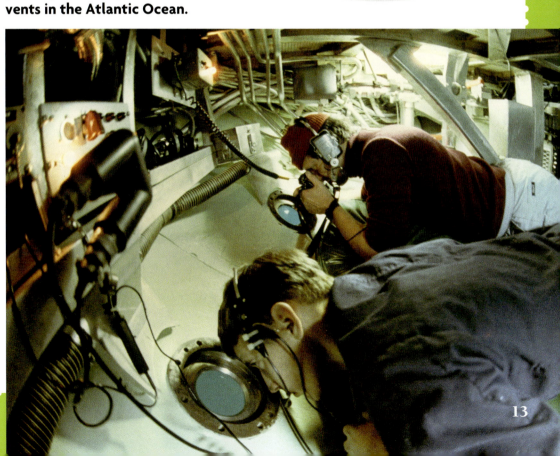

Reread the Informational Text

Analyze the Information in the Article
- What is the article mostly about?
- What is the main idea for each section of the article?
- How are the people in this article alike? How are they different?
- What conclusion(s) can you draw from this article about people who study the ocean?
- Which two pieces of information do you find most interesting? Explain your answer.

Focus on Comprehension: Make Inferences
- Why does Nancy Knowlton study coral reefs?
- What can you infer about people eating fish from polluted water?
- Exploring the deep can be dangerous. How can you tell?

Analyze the Tools Writers Use: A Strong Ending
Look at the last paragraph in this informational text (page 11).
- What type of ending is included in this informational text: a summary, restating important ideas, or making an observation?
- Does the ending make you rethink the topic? If yes, how? If no, why not?

Focus on Words: Word Origins

Make a chart like the one below. Read each root and meaning. Find and identify words from the text that were derived from that root. Look at both the root and how the word is used in the article to help you figure out what that word means. One example has been completed for you.

Page	Root and meaning	Word from text	Possible definition from root and text
7	*mar*; the sea	marine	of the sea
7	*organ*; tool		
11	*pion*; a foot soldier		
12	*phys*; nature		

coral colony in the Red Sea

What Lives in the Oceans?

The Census of Marine Life focused on three kinds of sea life. Above is a sea creature that swims freely.

Imagine living in a house that had one hundred rooms. But all your life you spent time in only five of those rooms. And though you knew about the other rooms, you had no idea what they looked like or what was in them. Wouldn't that be strange?

Actually, that has been scientists' experience of the world's oceans. The waters of Earth are vast. Yet only 5% of them have been explored. No one knows just how much life they contain—or how many kinds. In the year 2000, scientists decided to do something about this. They wanted to add to their knowledge. So they began the greatest ocean exploration in history. They decided to take notes, photos, and samples of what lives in the oceans. They called it the Census of Marine Life.

These sea creatures float or drift.

This sea creature lives on the ocean floor.

How the Census Was Done

The scientists had good reasons to count sea life. They knew that human actions were affecting the health of the oceans. Some sea creatures were beginning to die out. Others were increasing their populations very quickly. By finding out what was changing in the oceans, they might find ways to restore balance.

The **endeavor** was huge. More than 2,700 oceanographers and other scientists from more than eighty countries took part. They went on more than 500 ocean trips all over the world. Some went to the Arctic or Southern Oceans. Others explored the Pacific or Atlantic Oceans. And others went to the Indian Ocean.

They used all kinds of methods, including diving, fishing, and tracking with satellites. They studied the three kinds of life found in the oceans: creatures that live on the sea floor, creatures that swim freely, and creatures that float or drift on the surface. Of course, these scientists knew they couldn't count all the life in the seas. So they used a technique called sampling. They counted the life in one small part of an area. Then they estimated how much more life was living in a larger area.

The project cost about $650 million and took ten years to complete. The scientists wanted to spread the word about what they found, and so the results were published in at least fifty different languages.

What the Scientists Looked For

What did the Census find? To understand the answer, it's important to know what the scientists were looking for. They asked themselves three questions: What *lived* in the oceans? What *lives* in the oceans? What *will live* in the oceans?

Some scientists worked on answering the first question. They looked for bones or other signs of marine life that died. Other scientists focused on studying current forms of ocean life. They studied what the creatures look like and how they eat, move, and **reproduce**. Still other scientists **concentrated** on trying to predict the future of the oceans. Will some parts become too warm or too cold to sustain life? Will some have too much pollution for most sea life to survive?

What the Scientists Discovered

The scientists were amazed at what they found. According to an article in *Scientific American* magazine, they discovered more than 6,000 new kinds of sea creatures! For example, they discovered a spiny, many-legged animal the size of a speck of dust. They found it in both the Atlantic and Pacific Oceans. In the Indian Ocean, they found a snail with a layered shell unlike anything they had ever seen before.

leafy sea dragon

WHAT LIVES IN THE OCEANS?

And a new kind of jellyfish was discovered swimming in Arctic waters.

According to the Census study, scientists now know of about 250,000 different kinds of marine life. They also estimate that 70 to 80% of marine life is yet undiscovered. That means the total number of kinds of marine life may be as high as 1.4 million!

Napoleon wrasse fish

Scientists also concluded that three main human actions are threatening sea life:

1. Overfishing—When people catch too many of one kind of sea life, it reduces the number of those creatures. It also causes an imbalance in the environment.
2. Overdevelopment—Building homes along coasts or putting oil-drilling rigs out at sea can destroy the habitats, or places, where sea creatures live.
3. Pollution—People dumping trash into the oceans can kill fish that swallow it. And chemicals can poison the water.

Earth's ocean realm is a huge home to innumerable living things. It is their home, but part of ours, too. The Census of Marine Life provided a brief tour of this home. What other rooms are left to explore? What other surprises will scientists find as they open more doors?

How Do Scientists Explore the Oceans?

The Census scientists couldn't have done their work without special tools. This table shows some of the many **technologies** that oceanographers use.

Technology	Description	Advantages	Disadvantages
Divers	Humans wear special suits, carry tanks of air, and breathe through a tube connected to the tank.	Can explore very small spaces and make observations based on direct contact with rocks and living creatures.	Can't go more than about 140 feet deep, can't communicate directly with anyone on the surface, and aren't protected from predators. Can only stay underwater two hours at a time.
Submersibles	Underwater vehicles. Some are human driven, others are robots controlled by people on the ocean surface.	Can explore the deepest parts of the ocean and stay underwater for many hours. They can carry bulky cameras and other equipment. They can't be harmed by most ocean predators.	Robot submersibles are connected by cables to a surface ship and can't go very far. Dangerous storms can harm humans in submersibles.
Surface Vessels	Ships on the surface of the water.	Vessels can carry divers, diving gear, labs, computers, and even submersibles. Researchers can explore for months at a time.	Very expensive. Huge ships can travel only in large and deep areas.
Observational Tools	Airplanes and satellites that observe the oceans from miles above the surface. Tiny signal devices that are attached to marine animals.	Airplanes and satellites can observe and map huge areas that can't be seen from the ground, and can cover distances much faster than vessels. Rough weather poses no danger to satellites. Signal devices go everywhere a sea creature goes. They give information about that animal's behavior.	Airplanes and satellites are extremely expensive and offer no direct contact with ocean life. Signal devices run out of power, can stop working, or get lost.

Reread the Informational Text

Analyze the Information Presented in the Article
- What is the article mostly about?
- What is the main idea for each section of the article?
- In what ways do humans threaten sea life?
- What can you conclude from the chart on page 20?
- Which two pieces of information do you find most interesting? Explain your answer.

Focus on Comprehension: Make Inferences
- On page 17, the author says that scientists counted the life in one small part of an area. Then they estimated how much more life was living in a larger area. What can you infer about the accuracy of their estimate from this sentence?
- On page 19, the author says that scientists estimate that 70 to 80% of marine life is yet undiscovered. What can you infer about what scientists do know from this sentence?
- This text leads one to believe that one thing affects another. What evidence from the text supports this inference?

Analyze the Tools Writers Use: A Strong Ending
Look at the last paragraph in this informational text (page 19).
- What type of ending is included in this informational text: a summary, restating important ideas, or making an observation?
- How is this ending different than the ending in the first informational text?
- Does the ending make you rethink the topic? If yes, how? If no, why not?

Focus on Words: Word Origins
Make a chart like the one below. Read each root and meaning. Find and identify words from the text that were derived from that root. Look at both the root and how the word is used in the article to help you figure out what that word means.

Page	Root and meaning	Word from text	Possible definition from root and text
17	*deav*; duty		
18	*pro*; forward		
18	*centra*; center		
20	*tech*; art or skill		

Captain Moore

The author grabs readers' attention with a strong beginning: quotes from an interview with the man who first discovered the garbage patch. The first paragraph ends with an alarming piece of information that will keep readers interested in finding out more about the topic.

Ocean Garbage "Patches"

In 1997, Captain Charles Moore was sailing on the Pacific Ocean. Thousands of miles from shore, he noticed some pieces of plastic floating on the water. Then he saw more pieces. And more pieces. In a 2004 CBS News interview, he recalled that time. "Day after day, when I came on deck, I saw objects floating by: toothbrushes, bottle caps, and soap bottles," he said. He couldn't believe his eyes. What was going on? As it turned out, Moore discovered something huge—and disturbing. It was a floating garbage "patch" 1,000 miles west of California. This sea of garbage, which Moore described as "plastic soup," is now thousands of square miles across. Many scientists believe the Great Pacific Ocean Patch, as it is now called, is larger than the state of Texas!

If that wasn't disturbing enough, another garbage patch, nearly as large, was found on the Atlantic Ocean. This patch is 1,000 miles east of Bermuda. These areas aren't like garbage dumps on land. The trash isn't piled on top of itself. And no one could walk on it. Rather, a person on a boat might see an occasional plastic bag or bottle in the water. Most of the garbage has been turned into tiny bits by the force of winds and waves, and by hitting against other pieces of trash. Oceanographer Kara Lavender Law studies the Atlantic Ocean Patch. She told BBC News, "I think the word 'patch' can be misleading. This is widely **dispersed** . . . small pieces of plastic."

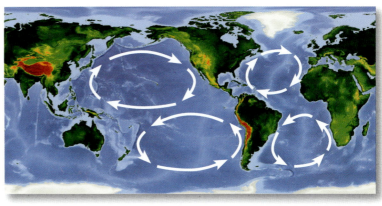

Ocean garbage "patches" form in convergence zones.

What Caused the Patches?

It may be hard to imagine such huge areas of trash floating in the middle of the ocean. Scientists, however, know that the answer to *how* these patches formed is easy. They are the result of two causes: humans and nature.

People often throw trash on the ground, into rivers, on the beach, or off boats. Rain and water flow carry that trash out to sea and into the ocean. Then nature takes over. Ocean winds and currents swirl and move the trash in different directions. But there are places in the ocean where the currents meet. In these places, called **convergence** zones, the winds and currents are weak. Large, calm areas are surrounded by great circles of water movement, called gyres (JY-erz). Much of the trash collects in these calm areas.

> The author organizes the information logically. The first paragraph of this section ends with a thesis sentence, or what she will talk about in the next paragraph. The following paragraph provides information that supports the thesis.

23

Ocean Garbage "Patches"

Sadly, the trash, most of which is made of plastic, can remain there for years and years. That is because plastic isn't **biodegradable**. It cannot be broken down easily by air, water, or sunlight. Fruit, paper, and cotton, on the other hand, are natural materials. They can dissolve in water in a matter of weeks or months. Man-made plastic takes decades or even centuries to break down. "I have no doubt that some of these things that we're discovering out there have been there since the dawn of the plastic era in the 1950s," said Moore.

How much trash is in the ocean? According to Greenpeace, an organization that fights to protect the environment, around 100 million tons of plastic are produced each year. The National Geographic Society believes the number is considerably higher, about 260 million tons. About ten percent of it ends up in the oceans.

> The author makes the information come alive for readers with a quote from Captain Moore.

> The author researched the topic in books and on the Internet, and got facts from different organizations.

How Long Does It Take Trash to Degrade?

orange or banana peel	2–5 weeks	newspaper	6 weeks
apple core	2 months	notebook paper	3 months
comic book	6 months	wool mitten	1 year
cardboard milk carton	5 years	wooden baseball bat	20 years
leather baseball glove	40 years	rubber sole of a shoe	50–80 years
steel can	100 years	aluminum soda can	350 years
plastic sandwich bag	400 years	plastic six-pack ring	450 years
glass bottle	1 million years	car tire	maybe never
polystyrene foam cup	maybe never		

Ocean Garbage "Patches"

Questions of Size

The bits of swirling plastic make it hard to tell exactly where the borders of an ocean patch are. The patch can't be seen by satellites or even airplanes. And, in fact, it's even hard to spot the floating plastic from a boat when the wind is blowing. "We don't know how big these patches are," said Law. "Nobody has actually defined the edges of where this is **accumulating**."

Some scientists, however, believe that the size of the patches has been exaggerated. Pollution expert Seba Sheavly is one of them. "Is there trash in the Pacific gyre? Of course there is. But is it the size of Texas? No, it's not," she said.

> The author presents multiple perspectives and differing opinions for readers to consider. (Note the author's use of the compare-contrast word "however.")

Oceanographer Angelicque White agrees with her. White has been studying the Pacific patch. She estimated that all of the floating trash put together would equal only one percent of the size of Texas. The difference in patch area depends on whether its measurement includes the spaces of water between the tiny bits of plastic.

> The author provides information. Readers draw the conclusions.

Whatever the actual size, patch-related problems worry scientists. About 70% of the floating trash eventually sinks to the bottom of the ocean. Some escapes the patch and floats into the rest of the ocean. At this point, there is no way to know just how much trash is in the water. All scientists can agree on is that there is way too much of it in the ocean. And that trash is harming sea life.

Ocean Garbage "Patches"

Dangerous Environmental Effects

Oceanographers and other scientists are alarmed by what they have discovered. According to the United Nations Environment Programme, more than a million seabirds and 100,000 marine mammals and turtles are killed every year by ocean litter. They eat pieces of plastic, thinking that it is food. Some of that plastic has soaked up oil and other poisons like a sponge. This poison is swallowed by sea life that eats the plastic. Marine animals also get caught in discarded nets or plastic rings from drink six-packs. They can choke or become trapped, unable to reach the surface to breathe.

The effects are also felt for thousands of miles beyond the patches. Microscopic organisms attach themselves to the plastic. Some crabs and sponges "hitch" rides on floating garbage. Currents can then carry them far away from their original locations. They settle in other habitats. They eat new prey and affect the balance of life in their new homes.

The author includes graphics that support the text.

OCEAN GARBAGE "PATCHES"

Lessons to Be Learned

Can anything be done to clean up the ocean garbage patches? Right now, scientists say no. By trying to scoop up the trash, people would also remove too many microscopic animals that are needed to maintain a balance in ocean life.

Scientists recommend that the best course of action is to prevent more trash from getting into the oceans. People should reduce the amount of trash they create. They should reuse objects rather than throw them out. And they should recycle trash so that it can be remade into other things.

As Ryan Yerkey of the Scripps Institute of Oceanography told BBC News, "Every piece of trash that is left on a beach or ends up in our rivers and washes out to the sea is an addition to the problem, so we need people to be the solution."

> The author concludes with a strong ending that quotes an expert in the field. This quote leaves readers thinking about what they might do to help solve the problem of ocean garbage patches.

Reread the Informational Text

Analyze the Information Presented in the Article
- What is the article mostly about?
- What is the main idea for each section of the article?
- What two viewpoints were discussed in this article?
- What do all of the scientists agree about?
- Which two pieces of information do you find most interesting? Explain your answer.

Focus on Comprehension: Make Inferences
- No one knows how many natural materials are put in the ocean. What evidence from the text supports this inference?
- The United Nations Environment Programme is able to distinguish how sea animals and birds die. What text evidence supports this inference?
- On page 27 the author says that nothing can be done to clean up ocean patches. What can you infer from this?

Focus on Quotation Marks
Quotation marks usually show direct speech, or someone speaking, in a story. However, quotation marks may also be used to set off text the author wants to emphasize. In this nonfiction text, the author uses quotation marks several times. On page 22, the author describes sea garbage as "plastic soup." We might not be able to clearly visualize sea garbage, but we certainly can visualize soup . . . with plastic in it. These words most closely describe what sea garbage looks like. Identify instances in this article where the author uses quotation marks to emphasize text. How is this feature helpful to you as a reader?

Analyze the Tools Writers Use: A Strong Ending

Look at the last paragraph in this informational text.
- What type of ending is used: a summary, restating important ideas, or making an observation?
- How is this ending different than the endings in the two previous informational texts?
- Does the ending make you rethink the topic? If yes, how? If no, why not?

Focus on Words: Word Origins

Make a chart like the one below. Read each root and meaning. Find and identify words from the text that were derived from that root. Look at both the root and how the word is used in the article to help you figure out what that word means.

Page	Root and meaning	Word from text	Possible definition from root and text
23	*disper*; to scatter		
23	*verge*; to lean forward		
24	*degrade*; down a step		
25	*cumul*; a heap		

THE WRITER'S CRAFT

How does an author write an
Informational Text?

Reread "Ocean Garbage 'Patches'" and think about what Jeanette Leardi did to write this informational text. How did she keep a narrow focus? How did she help you understand the text?

1. Decide on a Topic
Choose something you are interested in and want to know more about. Good writers enjoy researching their topics.

2. Narrow Your Focus
Jeanette Leardi knew she couldn't write about everything there was to know about the ocean, so she narrowed her focus to the recently discovered garbage "patches" found in the Pacific and Atlantic Oceans.

3. Write a Question About Your Focus
Questions lead to answers, so turn your focus into a question.

4. Research Your Focus
Become an "expert" by reading articles on the Internet, books, and newspaper articles, and by interviewing people connected with your topic. For instance, Jeanette Leardi listened to interviews with oceanographers studying garbage patches that originally aired on CBS News and the BBC.

5. Organize Your Information
Before writing an informational article, make a chart or table like the one on the next page that outlines the main points. For each main point, identify supporting details. You don't have to write full sentences. These are your notes. Remember, however, that there should be a logical progression of ideas.

6. Write Your Informational Text
As you write, develop each main point with your supporting details. Remember, you want people to enjoy reading your article as well as learn something new.

Topic: garbage "patches" found in the oceans
Focus: how they formed, problems they cause, what can be done
Question: What are ocean garbage "patches"?

Main Point	Details
Introduction: Discovery of Garbage "Patches" in Pacific and Atlantic Oceans	interview quotes from Capt. Moore with CBS News 2004, oceanographer Kara Law on BBC News
Causes	explain human contributions—trash on ground, in rivers, etc., goes to ocean; trash thrown overboard natural winds and currents (convergence zones, gyres; explain what they are); plastic not biodegradable; include sidebar chart of how long it takes for things to degrade include statistics about amount of plastic trash
Size: Conflicting Reports	quote and point of view from Law, who considers patches huge, "the size of Texas" quote from Seba Sheavly, environmentalist, that size is exaggerated; support with study results from Angelicque White that floating trash is only 1% size of Texas—if just the bits of plastic, not the area it covers in the water but both sides agree there is too much trash in water, 70% sinks to ocean floor
Problems	statistics from U.N. that a million seabirds and 100,000 marine mammals are killed by ocean litter annually explain about eating plastic that has absorbed oil and poisons; animals get caught in plastic rings from six-packs; sea creatures float to new areas and change balance of life
Conclusion	suggestions from scientists: can't clean up, but people need to put less trash in ocean; end with quote from Ryan Yerkey on BBC News

Glossary

accumulating (uh-KYOO-myuh-lay-ting) gathering together into a pile (page 25)

biodegradable (by-oh-dih-GRAY-duh-bul) able to be broken down into an energy source by the action of living things (page 24)

concentrated (KAHN-sen-tray-ted) focused effort on (page 18)

convergence (kun-VER-jens) the coming together of different parts into a whole (page 23)

dispersed (dih-SPERST) scattered; spread out (page 23)

endeavor (in-DEH-ver) a serious project usually requiring determined effort (page 17)

marine (muh-REEN) having to do with the sea (page 7)

organisms (OR-guh-nih-zumz) living things (page 7)

physical (FIH-zih-kul) having to do with natural science and what can be observed and measured (page 12)

pioneer (py-uh-NEER) a person who opens up a new way of thinking or activity (page 11)

reproduce (ree-pruh-DOOS) to produce offspring (page 18)

technologies (tek-NAH-luh-jeez) the practical applications of knowledge in certain areas (page 20)